BEI GRIN MACHT SICH IHR WISSEN BEZAHLT

- Wir veröffentlichen Ihre Hausarbeit,
  Bachelor- und Masterarbeit

- Ihr eigenes eBook und Buch -
  weltweit in allen wichtigen Shops

- Verdienen Sie an jedem Verkauf

Jetzt bei www.GRIN.com hochladen
und kostenlos publizieren

Stefanie Elzholz

Influence of shale gas development in Europe on gas market trends

GRIN Verlag

Impressum:

Copyright © 2011 GRIN Verlag GmbH
Druck und Bindung: Books on Demand GmbH, Norderstedt Germany
ISBN: 978-3-656-63528-4

Dieses Buch bei GRIN:

http://www.grin.com/de/e-book/271435/influence-of-shale-gas-development-in-
europe-on-gas-market-trends

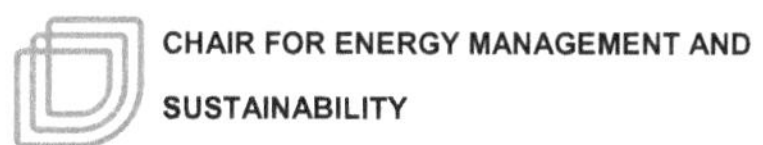

Seminar Report

Influence of shale gas development in Europe on gas market trends

by

Stefanie Elzholz

August 03, 2011

Universität Leipzig
Faculty of Economics and Business Management
Institute for Infrastructure and Resources Management (IIRM)
Vattenfall Europe Chair for Energy Management and Sustainability

Table of Contents

List of Figures

List of Abbreviations

Bcf/d	Billion cubic feet per day
CBM	Coal Bed Methane
CNG	Compressed Natural Gas
cp.	compare
Ed.	Editor
e. g.	exempli gratiā (for example)
eia	U. S. Energy Information Administration
f.	and the following one (page)
Fig.	Figure
GASH	Gas Shales in Europe (research project)
LNG	Liquefied Natural Gas
MMBtu	one million British thermal units
NGCC	Natural Gas Combined Cycle
n. n.	nomen nominandum (unknown author)
p.	page

Abstract

In Europe, a "new" unconventional energy resource begins to be explored. The so called *shale gas* isn't a new discovery, but its production has been too complex and expensive for a long time. Since rough quantities of shale gas have been exploited in North America, the global gas markets are changing. The European gas market is also affected by these changes which are characterized by an increasing gas supply and declining prices. Since Europe also has shale gas resources, the production of this unconventional gas might be attractive now. The present report investigates and evaluates the shale gas potentials in Europe. Against the background of energy policy targets like supply security and carbon emissions reduction, the option of shale gas as solution approach for these targets has to be evaluated. Although the production is meanwhile mostly profitable, active exploitation in Europe is not yet carried out because environmental concerns and the "green" lobby are countering. Nonetheless, shale gas is most likely to take a significant share of the future energy mix.

1 Introduction

Natural gas sources worldwide are sufficient to meet the primary energy demand over the coming five or six decades. Despite of this relatively optimistic forecast, Europe is strongly dependent on natural gas imports which can endanger the supply security in European countries. Russia and Iran carry the largest natural gas sources with a share of more than 40 percent of the worldwide total natural gas sources which leads to a geopolitical dependency. In addition to this, rising transportation costs and increasing primary energy consumption by China and India increase the uncertainty about a long-term availability of natural gas for Western Europe. To mitigate future supply bottlenecks, explorations of unconventional gas resources, the so called *shale gas*, could be a transitional solution. Shale gas already makes up ten percent of the current domestic gas production in the United States. In Europe there is also a potential but so far, there has been only little attention to this option. First exploration activities started currently now. (cp. SCHULZ/HORSFIELD 2010, p. 6)

In recent years, growing uncertainties, price volatility, rising demand and increasing costs pressurized the global energy industry. In order to reach and to maintain secure, sustainability and an affordable supply of energy, the focus has been shifting to alternative and promising resources like e. g. shale gas which seems to be abundant and almost globally available. (cp. WORLD ENERGY COUNCIL 2010, p. 3) Regardless to this assumption, activities in Europe concerning shale gas production are still reluctant due to the environmental concerns about the production process and associated political oppositions.

The present report aims to introduce shale gas as a potential "new" and unconventional fossil fuel in the European energy mix and to assess the resource concerning its future potential. After a brief introduction, fundamental basics about shale gas are presented in chapter two. Shale gas is initially defined and classified in the natural gas sector. In context with the production process, environmental concerns are discussed. Subsequently, the production process is described and worldwide shale gas sources and potentials are demonstrated. Chapter three is about the European gas market and how shale gas is developing in this country respectively how it is influencing its gas market. To this purpose, European sources and potentials are demonstrated and several ongoing projects are selected to be presented. In chapter four, the key findings of the report are summarized and a brief outlook is given.

2 Fundamental basics about shale gas

2.1 Definition and classification in the natural gas

Conventional natural gas as well as unconventional gas is generated in source rocks due to high heat and pressure. Conventional gas is stored in conventional respectively tight gas reservoirs as it is shown in figure 2-1 ("Conventional non-associated gas" and "Conventional associated gas"). In contrast, shale gas remains in place and is stored in nano-pores of the source rock. Its reservoirs, the so called gas shales, are deeper situated than conventional reservoirs as it is clearly seen in figure 2-1 ("Gas-rich shale"). (cp. DE VIVIÈS 2011, p. 3) Unconventional natural gases also include coal-bed methane (CBM), tight gas and methane hydrates. Gas shales are formations of organic rich shales, a sedimentary rock containing clay, quartz and other minerals. (cp. WORLD ENERGY COUNCIL 2010, p. 7). Differences between conventional and unconventional gas do not concern the usage but quality and composition of the gas that differ from shale to shale and from region to region due to geological particularities as it is known from conventional gas, too (cp. SCHULZ/HORSFIELD 2010, p. 6). The attractiveness of a shale gas resource compared to another is defined by its in-place concentration measured in Billion cubic feet per square mile (cp. EIA 2011b, p. 6).

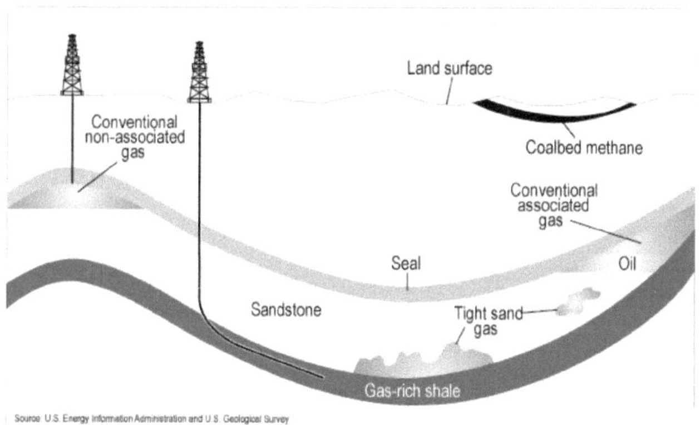

Fig. 2-1: The geology of natural gas resources (EIA 2011a)

2.2 Production process

The production process of shale gas is more complex than of conventional gas. Compared to conventional natural gas, shale gas cannot escape freely from the borehole. Therefore, the well runs horizontally once it reaches the gas-containing layer. In addition, water must be injected in order to produce cracks in the shale. Thus the rock becomes porous and the gas escapes better. (cp. RÜTH 2010) The used technologies are hydraulic fracturing to generate permeability and horizontal drilling to unlock tighter shale gas formations. They are combined with each other as two steps in one process. During the hydraulic fraction ("fracking") a fluid and a propping agent are pumped into the rock in order to fracture it. Equipment and technology of horizontal drilling is similar to those of vertical drilling. (cp. TYNDALL CENTRE 2011, p. 11 f.) The figure below demonstrates the process of exploitation.

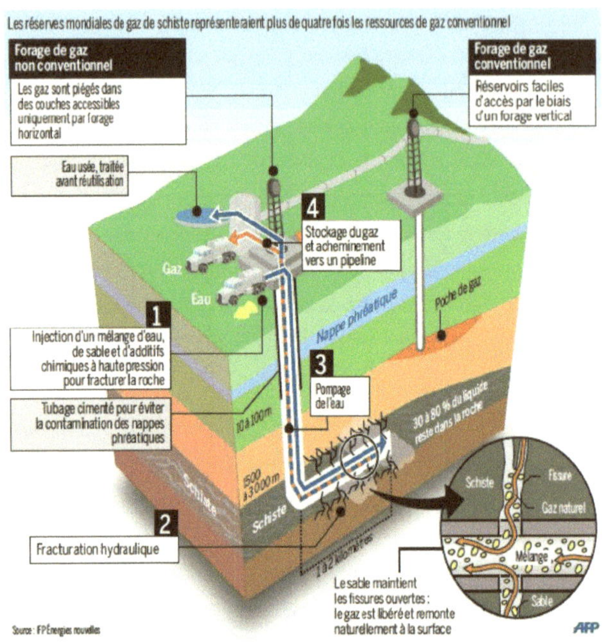

Fig. 2-2: Exploitation of shale gas (IFP ÉNERGIES NOUVELLES 2011)

The efficiency of the process depends greatly on the mineralogy of the shale, particularly its relative quartz, carbonate and clay content. Those with a relative high percentage of quartz and carbonate tend to be brittle, so that eliciting gas is easier compared to shales with relatively high clay content which tend to be ductile and to deform instead of shattering. A variety of further complex geologic features influencing the gas recovery efficiency exists. For further information, it is referred to the specified source of the U. S. Energy Information Administration (eia). (cp. EIA 2011b, p. 2-17 f.)

Against the background of the massive water demand during the drilling, water consumption represents another criterion in the assessment of the efficiency of the production process. Nevertheless, MANTELL claims that shale gas is one of the most water efficient raw fuels. Since it is domestically and regionally produced, additional environmental impacts of transporting the fuel as well as the net energy loss associated with transporting (fuel is needed for transport) are mitigated. This is why shale gas isn't just water efficient but also efficient from an energy perspective. The exploitation of shale gas requires water primarily during the drilling, but produces an enormous amount of natural gas over an approximate 20-year lifespan of the gas well. Compared to other fossil fuels, it is by far the most water efficient energy resource. Especially when used for power generation in a NGCC power plant, shale gas contributes to the most efficient ways of generating electricity. Furthermore, compressed natural gas is one of the most water efficient transportation fuels available today. (cp. MANTELL 2009, p. 13)

Despite of the high water and energy efficiency, shale gas exhibits economical disadvantages over the conventional gas production. The productivity of shale gas wells is lower and the recovery declines rapidly during the first couple of years. For compensating this drawback, a larger number of wells is necessary. With the requirement for horizontal drilling and hydraulic fracturing, shale gas wells are relatively expensive. The precise expensiveness technically depends on the shale characteristics as it is described above. Concerning the final costs, contrary opinions exist. While shale gas promoters claim costs of $5/MMBtu and below, skeptics estimate the costs at $7/MMBtu or more. Compared to the current gas price in Europe, this margin is between the long-term oil-linked contracts and the traded gas prices which makes shale gas quite competitive. (cp. GAS STRATEGIES 2010, p. 5)

To exploit shale gas reserves it will be necessary to develop an adequate infrastructure including a pipeline system for storage and distribution. Today, only a few of the proven basins possess an existing infrastructure. In the remaining basins, large investments are required which may result in delaying new production coming online or make the exploitation uneconomic. Nonetheless, each shale formation has to be evaluated on its specific merit. For strategic or other reasons, the formation may be still worth to exploit. (cp. WORLD ENERGY COUNCIL 2010, p. 3)

2.3 Environmental concerns

Due to the environmental concerns, shale gas production is controversial. The Energy Watch Group denounces the land use and above all the high water demand for the removal of shale gas. Provided economically feasible shale gas deposits in Europe, accordingly environmentally friendly mining techniques must be developed. (cp. RÜTH 2010) Another concern about shale gas production is the risk of groundwater contamination due to the chemicals used in hydraulic fracturing which are supposed to be carcinogenic and radioactive. However, experts bring forward the argument that the exploitation runs in such a deep, that the rock absorb the pollutants. (cp. RIA NOVOSTI 2011) TITZ also writes in his article, that most experts believe the risk of groundwater contamination is remote and cites the advantages for the climate. The combustion of gas produces less CO_2. In addition, gas-fired power plants can be readily combined with renewable energy sources. Furthermore, the former deposits of shale gas are potentially capable for carbon storage. (cp. TITZ 2010) Both the consumption of such large quantities of water and the risk of contamination of the groundwater are playing a larger role in the more densely-populated Europe than in the United States where shale gas is produced in remote areas. (cp. GAS STRATEGIES 2010, p. 5)

2.4 Sources and potential worldwide

Estimates quantify the worldwide shale gas deposits at more than 688 shales in 142 basins. At present, only a few dozen of these shales are exploited, most of them in North America (cp. WORLD ENERGY COUNCIL 2010, p. 3). The geological distribution of the basins is shown in figure 2-3. The global potential volumes of shale gas are estimated at about 453 trillion cubic meters (cp. SCHULZ/HORSFIELD 2010, p. 6). The current reserves of conventional natural gas are quantified at about 420 trillion cubic meters, that means, the worldwide gas supply could be doubled by the shale gas reserves (cp. RÜTH 2010).

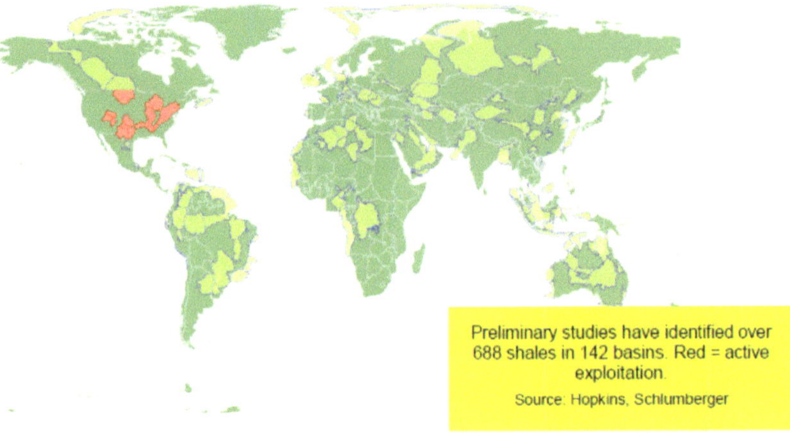

Fig. 2-3: Shale gas resources estimates worldwide (WORLD ENERGY COUNCIL 2010, p. 9)

As it is clearly seen in figure 2-3, North America still remains the only place with actual production. The United States of America can be described as the native country of shale gas production where it is already exploited for decades (cp. SCHULZ/HORSFIELD 2010, p. 7). Due to rising fuel prices and improved technologies, shale gas production became more and more profitable and competitive in recent years. Finally, shale gas displaced liquefied natural gas (LNG) from the U. S. gas market. (cp. TITZ 2010) However, no play outside North America has yet been proven and very variable contexts have to be considered, e. g. geological, industrial and environmental constraints (cp. DE VIVIÈS 2011, p. 7). Hence, the role of shale gas outside North America will play in future scenarios like it is demonstrated in the figures 2-4 and 2-5. According to the first graph, shale gas has a worldwide supply potential of 15 to 20 percent of the entire gas supply by 2040. The second graph shows that the USA, Canada and China will be the main players in future shale gas market.

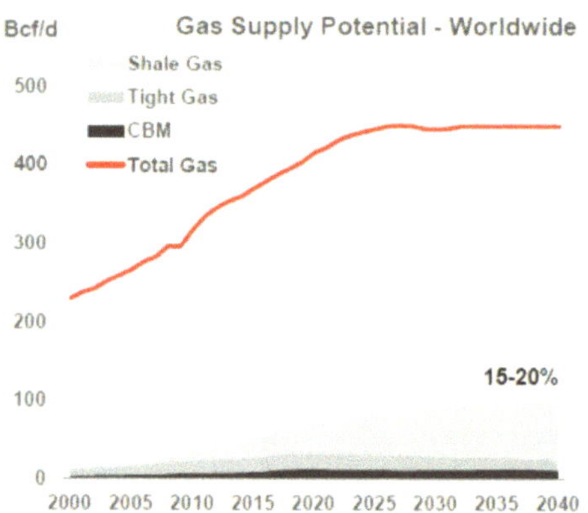

Fig. 2-4: Gas supply potential worldwide (DE VIVIÈS 2011, p. 10)

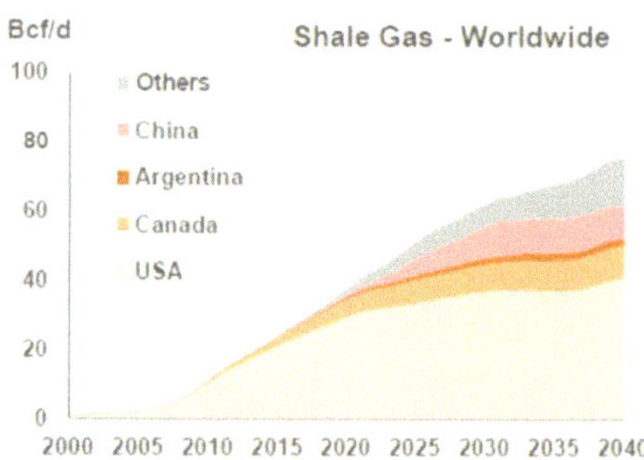

Fig. 2-5: Future shale gas development worldwide (DE VIVIÈS 2011, p. 10)

3 Shale gas developments in Europe

3.1 Overview over the European gas market

The European gas market is characterized by strong import dependence as it is also illustrated by the major trade flows of natural gas in figure 3-1. The main sources of supply of natural gas for Europe are Russia, Qatar and North Africa. It is transported by pipelines or in form of LNG by tankers.

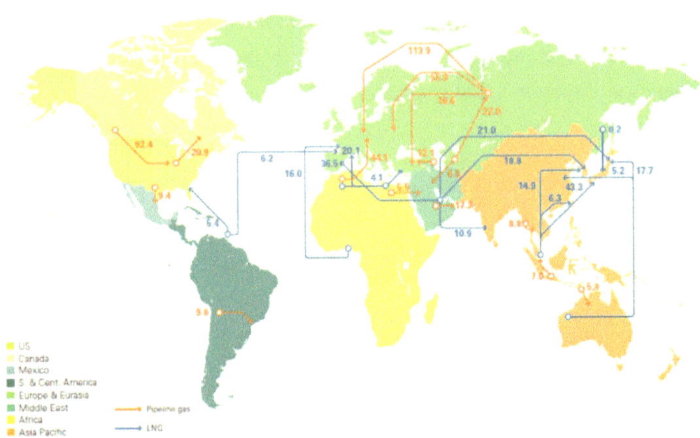

Fig. 3-1: Major trade flows of natural gas worldwide (billion cubic meters) (BP 2011b, p.29)

Due to long-term supply contracts and the oil price coupling, natural gas prices have been relatively high for many years. Until recently, the U. S. has been seen as the major prospective importer of natural gas, all above LNG. Though, the U. S. became self-sufficient owing to the growing shale gas sector. Consequently, LNG imports to the U. S. have been greatly scaled down and the oversupply gluts the European gas market where the prices come under pressure and decline. (cp. KRISTOFFERSEN 2011a) As it becomes apparent, the European gas market is already influenced by shale gas even though there are still no own exploitations.

Figure 3-2 demonstrates the future demand for natural gas in North America and Europe and by which types of natural gas it will be met. While unconventional gas has a growing importance for the American gas supply, Europe will first try to meet the growing demand by LNG and pipeline imports. Nonetheless, sooner or later, shale gas might become an attractive option for Europe to meet future demand. Technological advance and the efforts to reach supply security and import independence could be supportive for shale gas development in Europe.

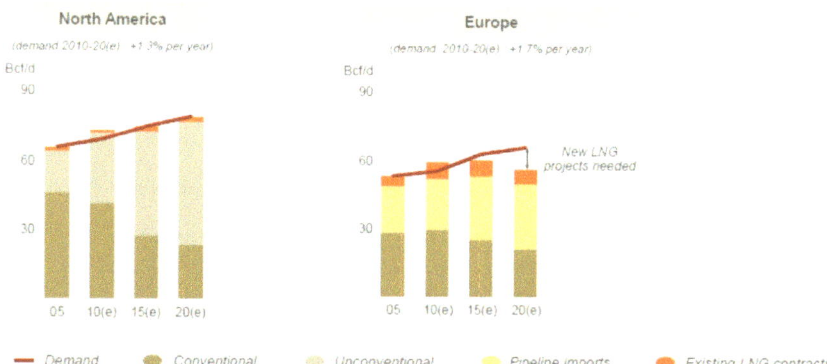

Fig. 3-2: Gas markets will tighten progressively in the coming years (DE VIVIÈS 2011, p. 2)

3.2 Sources and potential of shale gas

European shale gas reserves are estimated at 14 trillion cubic meters which almost corresponds to the same quantity of conventional natural gas resources (cp. RÜTH 2010). The main locations of shale gas reserves in Europe are illustrated in the map in figure 3-3.

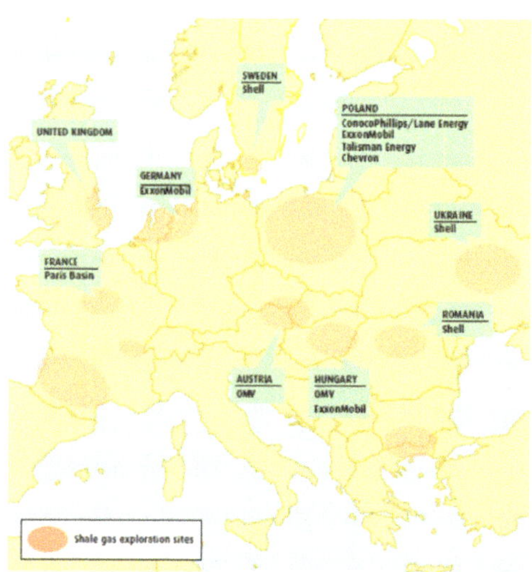

Fig. 3-3: Shale gas basins and exploration sites in Europe (GAS STRATEGIES 2010)

It becomes apparent that Poland carries the largest shale gas reservoirs in entire Europe why it is prospected to be the main player in future European shale gas market (fig. 3-4).

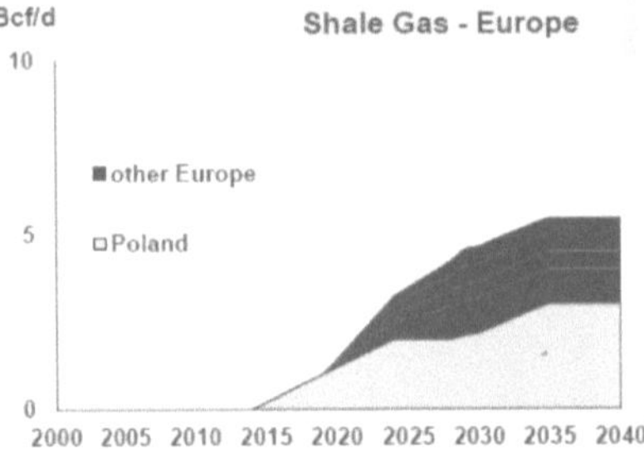

Fig. 3-4: Future shale gas development in Europe (DE VIVIÈS 2011, p. 10)

Despite the high potential, there are still no actual production activities anywhere in Europe. In Order to investigate the shale gas deposits, the research project Gas Shales in Europe (GASH) was founded in 2009. For these investigations, several exploration drillings have been carried out. It must be considered that European shales are not comparable to American shales. Europe had a very different geological history and thus, many potential reserves in Europe are less coherent. Furthermore, the U. S. has a tighter supply network and a highly deregulated gas market which are optimal conditions for the unconventional gas sector to develop. (cp. TITZ 2010)

Another reason for the sluggish development of shale gas exploitation in Europe are the resistances of the green energy lobby of the EU which wants to protect the fragile and heavily subsidized sector of green energy (cp. PEISER 2011).

3.3 Selected ongoing projects

3.3.1 Poland

There are already active levels of leasing and exploration in Polish shale gas basins. The main shales are deposited in three basins: The Baltic Basin in the north, the Lublin Basin in the south, and the Podlasie Basin in the east. They are supposed to have favorable characteristics for shale gas exploration. The mentioned three main basins are also illustrated in figure 3-5. Numerous large international and smaller exploration companies, as well as the country's national gas entity, are leasing actively the shales in the Baltic Basin. The most active company in the basin is 3Legs Resources, a subsidiary of Lane Energy Poland. In late September 2010, a joint venture between 3Legs and Conoco Phillips drilled the basin's first shale exploration wells, "Lebian LE1" and "Łęgowo LE1". Production information or other results have not yet been released. Talisman Energy plans to drill three further shale gas wells during the next two years. Furthermore, both Chevron and ExxonMobil have plans to drill exploration wells in 2011. As in the Baltic Basin, several international firms and the state owned gas company (PGNiG) are actively evaluating the shale gas potential of the Lublin Basin. In August 2010, Halliburton completed Poland's first shale gas well fracturing. Production and test results are still standing. A number of other exploration companies have acquired exploration concessions in the basin, including ExxonMobil, Chevron, Marathon Oil and others. No exploration drillings have yet been drilled into the shales of the Podlasie Basin but it is being actively leased. With three shale gas exploration concessions, ExxonMobil holds the largest lease position in the basin. (cp. EIA 2011b, p. V-1 - V-13) Shale gas exploration efforts in Poland purpose the objective to become more independent from Russian gas imports. The prospects are good: The domestic reserves are estimated at three trillion cubic meters minable shale gas. (cp. SCHULZ/HORSFIELD 2010, p. 7)

Fig. 3-5: Major shale gas basins of Poland (EIA 2011b, p. V-1)

3.3.2 Eastern Europe

Outside Poland, Eastern Europe also carries a couple of important basins containing promising shale gas resources but the potential has not yet been widely explored. The main basins are the Baltic Basin, the Lublin Basin and the Dnieper-Donets Basin as they are illustrated in the map in figure 3-6. Further resources are estimated in Hungary, Bulgaria and Rumania (Pannonian-Transylvanian Basin and Carpathian-Balkanian Basin). The shale gas potential of the Baltic Basin outside of Poland still has to be explored but leasing is not underway in the concerned countries. The investigations in the Dnieper-Donets Basin are against more advanced. EuroGas recently partnered with Total to explore the shale gas potential. First exploration drillings were carried out in 2010 but results have not yet been reported. The Lublin Basin outside Poland recently started to be explored by EuroGas in the Ukrainian part of the basin but there have not yet been any drillings. Exploration activities in the Pannonian-Transylvanian Basin are still in a speculative phase. East West Resources has been the first company to apply for concessions to explore the northern Rumanian part of the basin. In the Carpathian-Balkanian Basin, exploration drillings are carried out successfully by several firms since 2008. (cp. EIA 2011b, p. VI-1 - VI-24)

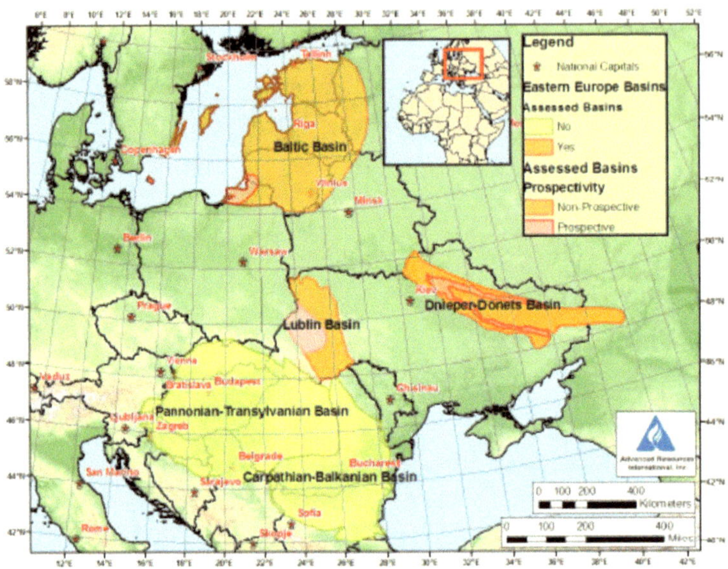

Fig. 3-6: Shale Gas Basins of Eastern Europe (EIA 2011b, p. VI-1)

3.3.3 Germany

The German territory covers the following basins: The North Sea-German Basin, the German Lower Saxony Basin, the Musterland Basin and the West Netherlands Basin. Ongoing exploration activities are still concentrated on the North-Sea-German Basin which is the largest one of the mentioned basins (see fig. 3-7). The leading company leasing prospective shale gas acreage in Germany is ExxonMobil which already drilled five exploration wells and plans an additional ten well exploration program in northwest Germany. Further companies that requested respectively already got concessions are BNK Petroleum and Realm Energy. (cp. EIA 2011b, p. VII-12 - VII-16) The exploration activities in Germany are critically evaluated and mainly not accepted by the public. In recent years, exploration drillings made headlines and numerous journals and broadcasts reported on the public resistance to the projects due to the suspected environmental and health risks. (cp. LOODEN 2010 or SCHULTZ 2010)

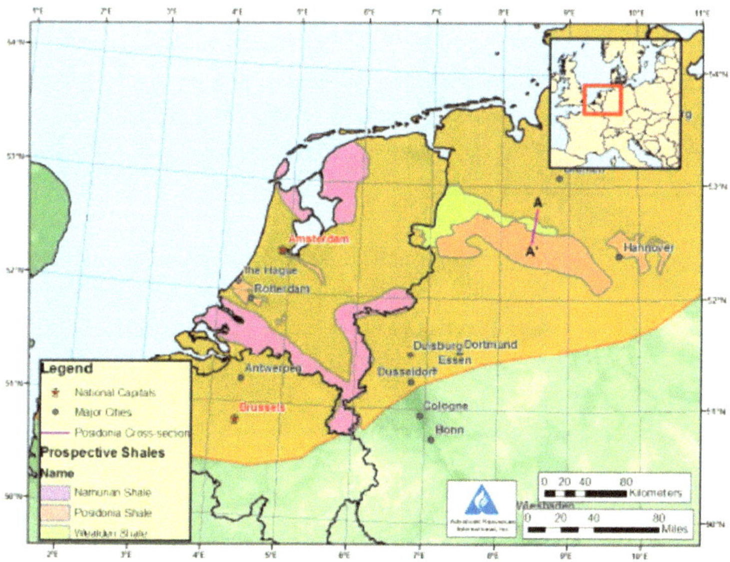

Fig. 3-7: North Sea-German Basin (EIA 2011b, p. VII-14)

3.4 Influence on gas market trends

As described in chapter 3.1, shale gas has already a significant influence on the European gas market due to the decreasing demand for gas imports in the U. S. as a consequence of an increasing domestic shale gas production over there. Due to the temporary oversupply of LNG in Europe, gas prices declined and set the entire market under pressure. As one consequence, even Gazprom has had to agree to deliver some of its gas at spot prices. Such incidents carry the potential both for existing as well as new players to develop and benefit from these evolutions. (cp. KRISTOFFERSEN 2011b)

It turned out that Europe also has a significant amount of shale gas but there is still uncertainty about the quality and the actual amounts why exploration activities are first necessary. As discussed in chapter 2.2, shale gas production is already quite efficient and competitive, all above in the U. S. where the required infrastructure is well developed. To reach a similar level in Europe, high investments are required but the rapid technological progress associated with increasing primary energy demand promises profitability. (cp. WORLD ENERGY COUNCIL 2010, p. 3; cp. DE VIVIÈS 2011, p. 2).

Contrary to renewable energy, shale gas developed without any support by subsidies or specific governmental measures. It is only driven by technological progress that makes shale exploration profitable. Due to the huge shale gas finds and the steadily increasing supply, gas prices declined sharply in recent years which benefits industry, households and energy security at the same time. Even though, shale gas itself does not play a major role in the current European primary energy mix, the rising importance of natural gas as bridging technology in the development from fossil fuels to renewable energy resources could incite shale gas exploration efforts. (cp. PEISER 2011)

A further consequence of declining gas prices is an expanding gas use so that coal use for power generation is gradually displaced by natural gas. In this way, the demand for primary energy is shifting from coal towards gas. The same mechanism can be observed in shifting oil demand, e. g. as transportation fuel, and a replacement through gas, e. g. as compressed natural gas (CNG) for transportation. (cp. WORLD ENERGY COUNCIL 2010, p. 5) Due to the long asset lifetimes, the fuel mix changes relatively slowly. After renewable energies but before other fossil fuels, natural gas experiences the fastest estimated growth by about 2.1 percent per year until 2030. (cp. BP 2011a, p. 17) As it becomes clear, the importance of natural gas for the primary energy mix will be strengthen due to promises of an increasing supply and a clean alternative to other fossil fuels.

In its Kuwait Report 2011, the Oxford Business Group claims that "domestic gas discoveries offer the prospect of a more diversified energy mix for Central European countries such as [...] for France and Germany, which import gas from all directions". (OXFORD BUSINESS GROUP 2011, p. 124) Becoming more independent on gas imports corresponds to one of the most important EU energy policy targets of supply security at reasonable prices. Even though, shale gas in Europe is at least one decade away, the outlook promises a significant development of shale gas in all major regions of proven reserves. The opportunity of a doubling of the natural gas supply is just too tempting against the background of rising demands and the challenge of clean energy. Figure 3-8 illustrates the development of sources of gas supply in North America, Europe and China. While shale gas is already important for the current North American gas mix, it will enter European and Chinese gas markets probably not till

2020 to 2030. LNG will also play a larger role in the future gas supply of European and Chinese gas market but it won't affect the U. S. gas market anylonger.

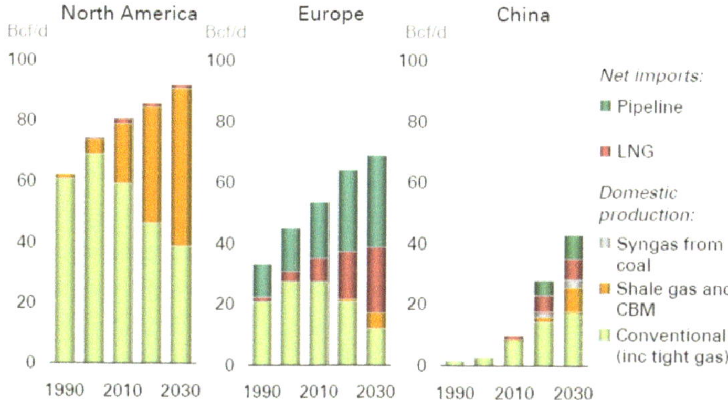

Fig. 3-8: Sources of gas supply in North America, Europe and China (BP 2011a, p. 54)

4 Conclusion

Shale gas remains a controversial energy resource containing lots of merits and risks. On the one hand, there are environmental impacts due to concerns about the used chemicals and due to the high water consumption. On the other hand, shale gas could contribute to increase independency on gas imports with the benefit of a higher supply security for the gas-importing countries.

Moreover, natural gas occupies a key role on the way to a sustainable energy market. It is the cleanest of hydrocarbons due to low carbon emissions, easy to control and efficient in distribution and use. It is the most efficient fossil fuel in conventional power generation and, due to its flexibility, natural gas is an ideal complement to renewable energy sources. (cp. IGU/EUROGAS 2010, p. 3-8). Especially against the background of the discredited nuclear energy, the role of natural gas as a bridging technology on the way to the age of renewables has strengthened. In this context, shale gas has to be taken into account as a real option.

To realize all the consequences of the increased production and use of shale gas, decades will pass. Numerous uncertainties can have a significant impact on the future of shale gas. Provided that there are still enough global reserves of conventional natural gas, there may not be sufficient incentive to exploit unconventional natural gas soon. It may also be the case that the production of unconventional gas is considerably more expensive than of conventional gas. (cp. WORLD ENERGY COUNCIL 2010, p. 5 f.) Therefore, the EU has to become clear about what cost it is willing to take to ensure its energy supply security and to reach more import independency compared to today (cp. KRISTOFFERSEN 2011b, p. 1).

In any case, the share of shale gas in the global energy mix is expected to develop and grow significantly. At least in 2030, shale gas is expected to be exploited actively in Europe to meet the growing demand. In return, the shares of coal and oil decline in favor of natural gas.

Not least, future potential of shale gas in Europe depends on creating public acceptance by acting responsibly. This includes a respectful interaction with local communities with respect to their interests as well as efforts to minimize the environmental impacts including protection of ground and groundwater plus minimizing water consumption. (cp. SCHULZ/HORSFIELD 2010, p. 8)

References

[BP 2011a] BP (Ed.) (2011): BP Energy Outlook 2030, in:
http://www.bp.com/liveassets/bp_internet/globalbp/globalbp_uk_english/reports_and_publications/statistical_energy_review_2011/STAGING/local_assets/pdf/2030_energy_outlook_booklet.pdf, 08/01/2011

[BP 2011b] BP (Ed.) (2011): BP Statistical Review of World Energy June 2011, in:
http://www.bp.com/assets/bp_internet/globalbp/globalbp_uk_english/reports_and_publications/statistical_energy_review_2011/STAGING/local_assets/pdf/statistical_review_of_world_energy_full_report_2011.pdf, 08/01/2011

[DE VIVIÈS 2011] DE VIVIÈS, Patrice (2011): Shale gas worldwide, Potential in Europe, at:
European Gas Conference June 2011

[EIA 2011a] EIA U.S. Energy Information Administration (Ed.) (2011): The geology of natural gas resources, in:
http://www.eia.gov/todayinenergy/images/NatGasSchematic.jpg, 06/20/2011

[EIA 2011b] EIA U.S. Energy Information Administration (Ed.) (2011): World Shale Gas Resources: An Initial Assessment of 14 Regions Outside the United States, in:
http://www.eia.gov/analysis/studies/worldshalegas/pdf/fullreport.pdf, 06/20/2011

[GAS STRATEGIES 2010] GAS STRATEGIES (Ed.) (2010): Shale gas in Europe: A revolution in the making? in:
http://www.gasstrategies.com/files/files/euro%20shale%20gas_final.pdf, 06/27/2011

[IFP ÉNERGIES NOUVELLES 2011] IFP ÉNERGIES NOUVELLES (Ed.) (2011): L'exploitation du gaz de schiste, in:
http://www.assemblee-nationale.fr/13/rap-info/i3517-4.gif, 08/01/2011

[IGU/EUROGAS 2010] IGU - INTERNATIONAL GAS UNION/EUROGAS - THE EUROPEAN UNION OF THE NATURAL GAS INDUSTRY (Ed.) (2010): The Role of Natural Gas in a Sustainable Energy Market, in:
http://www.eurogas.org/uploaded/The%20role%20of%20natural%20gas%20in%20a%20sustainable%20energy%20market%202010.pdf, 08/01/2011

[KRISTOFFERSEN 2011a] KRISTOFFERSEN, Stig-Arne (2011): A European Shale Gas Market: Fast developing or decades away? in:
http://www.articlesbase.com/industrial-articles/a-european-shale-gas-market-fast-developing-or-decades-away-4790841.html, 08/01/2011

[KRISTOFFERSEN 2011b] KRISTOFFERSEN, Stig-Arne (2011): Europe can lead the way with 'Green' Shale Gas development to make Shale Gas production a reality, in:
http://gasgefluester.de/2011/05/europe-can-lead-the-way-with-green-shale-gas-development-to-make-shale-gas-production-a-reality/?page=1, 08/01/2011

[LOODEN 2010] LOODEN, Silke (2010): Erdgassuche mit giftiger Bohrflüssigkeit, in: Weser Kurier 11/11/2010
http://www.weser-kurier.de/Druckansicht/Region/Niedersachsen/263296/Erdgassuche%20+mit+_giftiger+Bohrfluessigkeit.html, 07/02/2011

[MANTELL 2009] MANTELL, Matthew E. (2009): Deep Shale Natural Gas: Abundant, Afforda
ble, and Surprisingly Water Efficient, in:
http://www.energyindepth.org/wp-
content/uploads/2009/03/MMantell_GWPC_Water_Energy_Paper_Final.pdf,
08/01/2011

[OXFORD BUSINESS GROUP 2011] N. N. (2011): Shale Storm, in: (Oxford Business Group) The
Report: Kuwait 2011, p. 123-125

[PEISER 2011] PEISER, Benny (2011): Gold rush for shale gas or false dawn? in:
http://www.publicserviceeurope.com/article/317/gold-rush-for-shale-gas-or-false-
dawn, 08/01/2011

[RIA NOVOSTI 2011] N. N. (2011): Schiefergas-Boom in Europa: Verliert Gazprom an Boden?
- „Nesawissimaja Gaseta", in: (RIA Novosti)
http://de.rian.ru/industry_agriculture/20110111/258063885.html, 08/01/2011

[RÜTH 2010] RÜTH, Christine (2010): Gasreserven in Tongestein, Max-Planck-Institut für
Plasmaphysik, in:
http://www.ipp.mpg.de/ippcms/ep/ausgaben/ep201002/0210_shalegas.html,
06/12/2011

[SCHULTZ 2010] SCHULTZ, Stefan (2010): Gasbohrung - US-Konzern presste giftige Chemika-
lien in Niedersachsens Boden, in: Spiegel Online
http://www.spiegel.de/wirtschaft/unternehmen/0,1518,725697,00.html,
07/02/2011

[SCHULZ/HORSFIELD 2010] SCHULZ, Hans-Martin/HORSFIELD, Brian (2010): Shale Gas: New
Unconventional Gas Resource for Europe, in: ew - das magazin für die energie
wirtschaft, 01/2010, p. 6-8

[TITZ 2010] TITZ, Sven (2010): Schiefergas - Die wiederentdeckte Reserve, Neue Techniken
ermöglichen die rentable Ausbeute unkonventioneller Gasvorkommen, in:
http://www.nzz.ch/nachrichten/hintergrund/wissenschaft/klimawandel/die_wissen
schaftli-
che_sicht_der_dinge/schiefergas__die_wiederentdeckte_reserve_1.4448827.htm
l, 05/30/2011

[TYNDALL CENTRE 2011] TYNDALL CENTRE FOR CLIMATE CHANGE RESEARCH (Ed.) (2011):
Shale gas: a provisional assessment of climate change and environmental impacts, in:
http://www.tyndall.ac.uk/sites/default/files/tyndall-
coop_shale_gas_report_final.pdf, 07/29/2011

[WORLD ENERGY COUNCIL 2010] WORLD ENERGY COUNCIL (Ed.) (2010): Survey of Energy Resources: Focus on Shale Gas, in:
http://www.worldenergy.org/documents/shalegasreport.pdf, 06/27/2011